STUDY GUIDES

Mathematics

Year 3

Jenny Lawson

RISING STARS

Rising Stars UK Ltd, 7 Hatchers Mews, Bermondsey Street, London SE1 3GS

www.risingstars-uk.com

All facts are correct at time of going to press.

Published 2007
Reprinted 2010, 2011, 2012
Text, design and layout © Rising Stars UK Ltd.

Design: HL Studios and Clive Sutherland
Illustrations: Oxford Designers and Illustrators
Editorial project management: Dodi Beardshaw
Editorial: Jane Wilsher
Cover design: Burville-Riley Partnership

British Library Cataloguing in Publication Data.
A CIP record for this book is available from the British Library.

ISBN: 978-1-84680-093-1

Printed in Great Britain by Ashford Colour Press Ltd, Gosport, Hampshire

Contents

How to use this book

How to get the best out of this book   4

Counting and understanding number

1   Numbers up to 1000   6
2   Partitioning 3-digit numbers   8
3   Rounding and estimating   10
4   Proper fractions   12

Knowing and using number facts

5   Addition facts   14
6   Subtraction facts   16
7   Multiplication facts   18
8   Division facts   20
9   Double checking   22

Calculating

10   Mental addition and subtraction   24
11   Written addition   26
12   Written subtraction   28
13   Multiplying by 10 or 100   30
14   Written multiplication   32
15   Written division   34
16   Division as the inverse of multiplication   36
17   Unit fractions   38

Understanding shape

18   2-D and 3-D shapes   40
19   Reflections in a mirror   42
20   Position, direction and movement   44
21   Right angles in 2-D shapes   46

Measuring

22   Units of length, weight and capacity   48
23   Reading scales   50
24   Intervals of time   52

Handling data

25   Tally charts and frequency tables   54
26   Using diagrams to sort data   56

Using and applying mathematics

27   Problem solving   58
28   Using images and diagrams to solve problems   60
29   Following lines of enquiry   62
30   Patterns   64

How to get the best out of this book

Most chapters spread across two pages but some spread over four pages. All chapters focus on one topic and should help you to keep 'On track' and to 'Aim higher'.

Title: tells you the topic for the chapter.

What do you need to know? and **What will you learn?** tell you what you need to know before you start this chapter and what you are aiming to learn from this chapter.

Key facts: set out what you need to know and the ideas you need to understand fully.

Key words and their meanings: help to build up your mathematical vocabulary. Remember that some words mean one thing in everyday life and something more special in Mathematics.

Year **3**   Understanding shape

18 2-D and 3-D shapes

What do you need to know?

- What common 2-D shapes and 3-D solids look like in different positions and orientations
- Sort, make and describe 2-D shapes, referring to their properties

What will you learn?

- How to describe and classify 3-D solids
- How to draw and make 2-D shapes

Example

- To draw a 2-D shape, first check how many sides it has. Note any special properties, such as right angles, equal sides or parallel lines. Use a ruler to draw the straight lines.

Draw shapes with right angles on squared paper.

Draw equilateral triangles and hexagons on an isometric grid.

Key facts

A **triangle** has three angles. A triangle with all three sides the same is called **equilateral**. A triangle with only two sides the same length is called **isosceles**. If all the sides are different lengths the triangle is called **scalene**.

Equilateral

Isosceles

Scalene

A **quadrilateral** has four sides. **A square** is the quadrilateral with all sides the same length and all angles the same. Curved lines can be used to create shapes: **semi circles**, **circles** and **ovals**.

Square

Circle

Semi circle

Oval

Language

2-D shapes are made from **lines**. The point where any two lines meet to form a corner is called a **vertex**. **3-D solids** are constructed from shapes, called **faces**, or sides. Pairs of faces join along a solid **edge**.

40

Follow these simple rules if you are using the book for revising.

1 Read each page carefully. Give yourself time to take in each idea.

2 Learn the key facts and ideas. Ask your teacher or mum, dad or the adult who looks after you if you need help.

3 Concentrate on the things you find more difficult.

4 Only work for about 20 minutes or so at a time. Take a break and then do more work.

If you get most of the **On track** questions right then you know you are working at level 3 or 4. Well done – that's brilliant! If you get most of the **Aiming high** questions right, you are working at the higher level 4. You're doing really well!.

The **Using and applying questions** are often more challenging and ask you to explain your answers or think of different ways of answering. These questions will be around level 4 or above.

Some questions must be answered without using a calculator – look for ![calculator crossed out]. If you are not using a calculator be sure to write down the calculations you are doing. If you are using a calculator remember to try to check your answer to see if it is sensible.

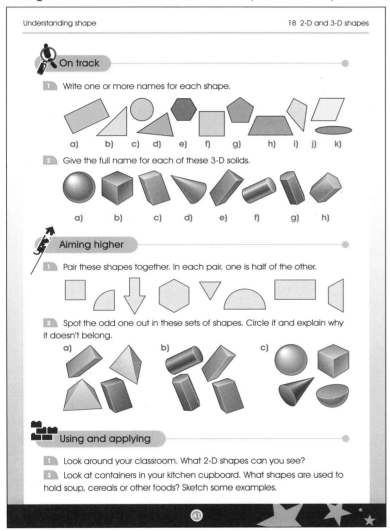

Follow these simple rules if you want to know how well you are doing.

1 Work through the questions.

2 Keep a record of how well you do.

3 If you are working at level 3 you will get most of the On track questions correct.

4 If you are working at level 4 you will also get most of the Aiming higher questions correct.

1 Numbers up to 1000

What do you need to know?

- How to read and write 2-digit and 3-digit numbers in figures and words
- How to order 2-digit numbers and position them on a number line

What will you learn?

- How to read, write and order whole numbers to at least 1000
- How to position numbers up to 1000 on a number line
- How to count on from zero and back to zero in single-digit steps or multiples of 10

Example

- The number after 9 is 10. In the number 10, the digit 1 stands for 1 ten. The 0 is a placeholder. It says there are no units.
- The higher you count, the bigger the numbers grow and they need more digits.

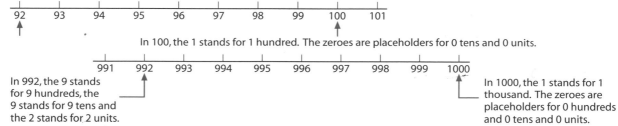

In 100, the 1 stands for 1 hundred. The zeroes are placeholders for 0 tens and 0 units.

In 992, the 9 stands for 9 hundreds, the 9 stands for 9 tens and the 2 stands for 2 units.

In 1000, the 1 stands for 1 thousand. The zeroes are placeholders for 0 hundreds and 0 tens and 0 units.

Key facts

The number after 99 is 100. The number after 999 is 1000.

Language

2-digit number: the first digit is the number of tens and the second digit is the number of units.

3-digit number: the first digit is the number of hundreds, the second digit is the number of tens and the third digit is the number of units.

Place value: this is the position of a digit in a number, such as first, second or third, which shows the number of hundreds, tens or units.

On track

1. Write these numbers in figures.
   a) two hundred and forty-five
   b) six hundred and eight

2. Complete this number line by writing in the missing numbers.

   8   9   11   14   18

3. Put these numbers in order, starting with the smallest.

   37   65   15   73   56   51   89   98

Aiming higher

1.
   0 1 2 3 4 5 6 7 8 9 10 11 12 13 14 15

   a) What number is 3 more than 1?
   b) What number is 2 less than 5?
   c) What number is halfway between 2 and 10?

2.
   95   96   97   98   99   100   101   102   103   104   105

   a) What number is 3 more than 98?
   b) What number is 4 less than 103?
   c) What number is halfway between 95 and 105?

3. Put these numbers in order, starting with the largest.

   570   715   575   705   750   157   751   515

Using and applying

1. The houses on Kate's side of Acacia Avenue have even numbers. Kate lives at number 34 and Lucy lives three doors down. What number house does Lucy live at?

2. On 10 February, at noon, the temperature was 14°C. At night, the temperature dropped by 5°. How cold was it at night?

2 Partitioning 3-digit numbers

What do you need to know?

- What each digit in a 2-digit number stands for
- How zero can be a placeholder
- How to partition 2-digit numbers into multiples of 10 and 1

What will you learn?

- How to partition 3-digit numbers into multiples of 1, 10 and 100

Example

- To do a subtraction such as 572 – 241 each 3-digit number can be partitioned.

    $572 - 241 \quad = 500 + 70 + 2 - 200 - 40 - 1 = 300 + 30 + 1 = 331$

- For subtractions like 572 – 349, you might partition 572 differently.

    $572 - 349 = 500 + 60 + 12 - 300 - 40 - 9$

    $= 500 - 300 + 60 - 40 + 12 - 9$

    $= 200 + 20 + 3$

    $= 223$

Key facts

In a 3-digit number, such as 239, the 2 represents 2 hundreds, the 3 represents 3 tens and the 9 represents 9 units.

H T U	H T U	H T U
2 3 9	2 4 0	3 0 7

A zero is a **placeholder**. In 240, there are hundreds and tens but no units. In the number 307, there are hundreds and units but no tens.

Language

Placeholder: a zero that shows there are none for that column.
Partitioned: when a number is broken into smaller numbers. This can be done in many ways, e.g. 57 = 50 + 7 = 40 + 17.

On track

1　What does the 4 stand for in each of these numbers?

a)　45　　　　　b)　64　　　　　c)　401　　　　　d)　143

2　Partition these numbers into tens and units.

a)　36　　　　　b)　73　　　　　c)　41　　　　　d)　82

3　Partition these numbers into hundreds, tens and units.

a)　136　　　　b)　291　　　　c)　372　　　　d)　425

Aiming higher

1　Partition these numbers in more than one way.

a)　245　　　　b)　387　　　　c)　439　　　　d)　521

2　With the digits 1, 6 and 9, you can make six different 3-digit numbers.

a)　What is the value of the digit 1 in each of these 3-digit numbers?

169　　　　196　　　　619　　　　691　　　　916　　　　961

b)　Arrange the digits 2, 4 and 6 to make two 3-digit numbers, each with the 6 representing 60.

Using and applying

1　Amy is counting on in tens: 27, 37, 47, 57, ...

26	27	28
36	37	38
46	47	48
56	57	58

Carry on counting on in tens until you go past 100.

2　Amy is counting back in hundreds: 854, 754, 654, ...

Carry on counting back in hundreds until you go past 100.

3　Make up your own counting on or counting back puzzle and try it out on a friend.

3 Rounding and estimating

What do you need to know?

- How to estimate a number of objects
- How to round 2-digit numbers to the nearest 10

What will you learn?

- How to round 3-digit numbers to the nearest 10 or 100
- How to estimate sums and differences of numbers

Example

- Estimate the sum of 46 and 112.

46	Rounding 46 to the nearest 10 is 50.	50
112	Rounding 112 to the nearest 10 is 110.	110
158	Adding 50 and 110 gives 160.	160

Check: 46 + 112 = 158. Notice that 158 to the nearest 10 is 160.

Key facts

When you round to the nearest 10, the rounded number will have a 0 in the unit column. To decide what goes in the tens column, look at the units figure.

| TU | | TU |
| 37 | ⇑ | 40 | If the units figure is 5 or above, round up.
| 34 | ⇓ | 30 | If the units figure is less than 5, round down.

When you round to the nearest 100, to decide what goes in the hundreds column, look at the tens figure.

Language

Rounding takes a number to the nearest 10 or the nearest 100.
Rounding up takes a number up to the next 10 or 100.
Rounding down takes a number down to the previous 10 or 100.

On track

1. Round these numbers to the nearest 10.
   a) 62   b) 76   c) 29   d) 44   e) 35   f) 51

2. Round these numbers to the nearest 10.
   a) 519   b) 861   c) 229   d) 673   e) 333   f) 749

3. Round these numbers to the nearest 100.
   a) 519   b) 861   c) 229   d) 673   e) 333   f) 749

4. Round your answers to question 2 to the nearest 100.
Compare these rounded numbers with your answers to question 3.

Aiming higher

1. Estimate the answers to these sums by rounding the numbers to the nearest 10.
   a) 27 + 32   b) 49 + 53   c) 84 + 17   d) 61 + 75

2. Estimate the answers to these sums by rounding the numbers to the nearest 100.
   a) 428 + 337   b) 142 + 855   c) 284 + 717   d) 561 + 275

3. Work out the exact answers for questions 1 and 2. How close were your approximated answers?

Using and applying

1. Katie's birthday is on 1st September. Estimate to the nearest 10 how many days there are between 1st January and her birthday. Give your answer to the nearest 10 days.

2. A cinema has 12 rows of seats. Each row has 19 seats. Approximately how many seats are there in the cinema? Give your answer to the nearest 10 seats.

3. Work out the exact answers for questions 1 and 2. How close were your estimates?

4 Proper fractions

What do you need to know?

- How to find one half, one quarter and three quarters of a shape

What will you learn?

- That in a fraction such as $\frac{3}{8}$, the 3 means 3 parts of a whole and the 8 means the total number of parts
- How to identify and estimate fractions of shapes
- How to use diagrams to compare fractions and to find equivalent fractions

Example

- Each of these circles has been divided into 8 slices and 3 slices have been shaded. Notice that $\frac{3}{8}$ of a circle can be shaded in different ways.

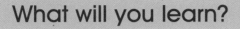

- In this circle, $\frac{4}{8}$ of the circle has been shaded. This is the same as shading $\frac{2}{4}$ of the circle or $\frac{1}{2}$ of the circle.

Key facts

If the **numerator** is smaller than the **denominator**, the fraction is less than 1. If the numerator is the same as the denominator, the fraction is equivalent to 1.

$$\frac{2}{2} \quad = \quad \frac{4}{4} \quad = \quad \frac{8}{8} = 1$$

Equivalent fractions give the same "share", e.g. $\frac{2}{4} = \frac{4}{8}$.

Language

Numerator: the number on the top of a fraction. $\longrightarrow \frac{4}{5}$
Denominator: the number on the bottom of a fraction. $\longrightarrow \frac{4}{5}$

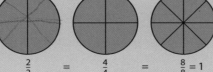

 On track

1 Which of these fractions is less than 1?

$\frac{1}{7}$ $\frac{5}{9}$ $\frac{9}{5}$ $\frac{7}{1}$

2 What fraction of each of these squares has been coloured?

3 What fraction of this circle has been shaded?

4 Shade $\frac{5}{8}$ of this circle.

 Aiming higher

1 In these circles, what fraction has been shaded?

3 In these squares, what fraction has been shaded?

2 Complete this statement:

$\frac{1}{3} = \frac{?}{6} = \frac{3}{9}$

4 Complete this statement:

$\frac{2}{3} = \frac{4}{?} = \frac{6}{9}$

Using and applying

1 Draw two circles, one divided into four equal segments and one divided into eight equal segments. Shade them to show that $\frac{1}{4} = \frac{2}{8}$.

2 Draw two circles, each divided into four equal segments. Shade an area to show the fraction $\frac{5}{4}$.

5 Addition facts

What do you need to know?

- All addition facts for each number to at least 10
- All number pairs that total 20
- All pairs of multiples of 10 with totals up to 100

What will you learn?

- All addition facts for each number to 20
- All sums of multiples of 10
- All number pairs that total 100

Example

- You can use your knowledge of addition facts for each number to 10 to learn the facts for numbers to 20.

$$10 = 1 + 9 = 2 + 8 = 3 + 7 = 4 + 6 =$$
$$5 + 5 = 6 + 4 = 7 + 3 = 8 + 2 = 9 + 1$$

11 is 1 more than 10.

$$11 = 1 + 10 \qquad then \qquad 11 = 2 + 9 = 3 + 8 = 4 + 7 = 5 + 6 = \dots$$

- For 20, you know that $10 + 10 = 20$.

$$20 = 1 + 19 = 2 + 18 + 3 + 17 = 4 + 16 = 5 + 15$$
$$= 6 + 14 = 7 + 13 = 8 + 12 = 9 + 11$$

Key facts

Use your knowledge of number bonds:
$$10 = 1 + 9 = 2 + 8 = 3 + 7 = 4 + 6 = \dots$$
To add multiples of 10: $100 = 10 + 90 = 20 + 80 = 30 + 70 = \dots$

Language

Number pair: any two numbers that have a special property, such as adding up to 100. For example:
$$51 + 49 = 100 \qquad 37 + 63 = 100 \qquad 18 + 82 = 100 \qquad 24 + 76 = 100$$
Notice that the tens digits add to 9 and the units digits add to 10.

On track

1 What are the missing numbers?

   a) 11 + ? = 20   **b)** 13 + ? = 20

   c) 15 + ? = 20   **d)** 17 + ? = 20

2 Copy and complete this addition table for numbers 11 to 20.

+	11	12	13	14	15	16	17	18	19	20
11	22	23	24	25	26	27	28	29	30	31
12	23	24	25	26	27	28	29	30	31	32
13	24	25	26	27	28	29	30	31	32	33
14	25	26	27	28	29	30	31	32	33	34
15	26	27	28	29	30	31	32	33	34	35
16	27	28	29	30	31	32	33	34	35	36
17	28	29	30	31	32	33	34	35	36	37
18	29	30	31	32	33	34	35	36	37	38
19	30	31	32	33	34	35	36	37	38	39
20	31	32	33	34	35	36	37	38	39	40

Compare your table with the addition table for the numbers 1 to 10.

Aiming higher

1 What are the missing numbers?

   a) 41 + ? = 100   **b)** 73 + ? = 100   **c)** 35 + ? = 100

   d) 67 + ? = 100

2 Complete these sums.

   a) 30 + 50 =   **b)** 20 + 60 =   **c)** 40 + 10 =

   d) 20 + 30 =   **e)** 60 + 10 =   **f)** 20 + 20 =

Using and applying

1 Add across the rows and add down the columns. Check that your totals in the third row and third column agree.

6	+	5	⇒	11
+		+		+
2	+	7	⇒	9
⇓		⇓		⇓
8	+	12	⇒	20

6 Subtraction facts

What do you need to know?

- All subtraction facts for each number to at least 10
- All pairs of numbers with totals to 20
- All pairs of multiples of 10 with totals up to 100

What will you learn?

- All subtraction facts for each number to 20
- All differences of multiples of 10

Example

- You can use your knowledge of subtraction facts for each number to 10 to learn the facts for numbers to 20.

  $10 - 1 = 9$ $20 - 1 = 19$

  $10 - 2 = 8$ $20 - 2 = 18$

  $10 - 3 = 7$ $20 - 3 = 17$

- You can also use your knowledge of subtractions facts such as:

  $10 - 9 = 1$ $10 - 8 = 2$ $10 - 7 = 3$ $10 - 6 = 4$

  to help you to do subtractions of multiples of 10.

  $100 - 90 = 10$ $100 - 80 = 20$ $100 - 70 = 30$ $100 - 60 = 40$

Key facts

For every addition fact, there are two subtraction facts.

$9 + 1 = 10$	$10 - 1 = 9$	$10 - 9 = 1$
$8 + 2 = 10$	$10 - 2 = 8$	$10 - 8 = 2$

Language

Multiple of a number: what you get when you multiply it by another number. 10 and 15 are both multiples of 5.

On track

1 What are the missing numbers?

    a) $10 - ? = 4$ **b)** $10 - ? = 2$ **c)** $10 - ? = 6$ **d)** $10 - ? = 8$

Write an addition fact that helped you to work out each answer.

2 What are the missing numbers?

    a) $70 - 10 = ?$ **b)** $90 - 20 = ?$ **c)** $80 - 20 = ?$ **d)** $60 - 10 = ?$

Write an addition fact that helped you to work out each answer.

Aiming higher

1 What are the missing numbers?

    a) $20 - ? = 4$ **b)** $20 - ? = 2$ **c)** $20 - ? = 6$ **d)** $20 - ? = 8$

Write an addition fact that helped you to work out each answer.

2 $11 + 4 = 15$ so $15 - 4 = 11$ and $15 - 11 = 4$

Choose two other numbers that add up to 15 and write an addition fact. Then write two subtraction facts using the number 15 and your two chosen numbers.

3 Repeat question 2, but for pairs of numbers which add up to

    a) 16   **b)** 17   **c)** 18   **d)** 19

4 What are the missing numbers?

    a) $100 - ? = 43$ **b)** $100 - ? = 27$ **c)** $100 - ? = 62$ **d)** $100 - ? = 89$

Write an addition fact that helped you to work out each answer.

Using and applying

1 In this number square, the sum of the first two cells in each row equals the number in the third cell. The same is true for the columns. Use addition and subtraction facts to complete the square.

16 +	28 ⇒	44
+	+	+
25 +	31 ⇒	56
⇓	⇓	⇓
41 +	59 ⇒	100

2 Make up your own square with the corner number of 100.

7 Multiplication facts

What do you need to know?

- All multiplication facts for the 2, 5 and 10 times tables
- How to recognise multiples of 2, 5 or 10

What will you learn?

- All multiplication facts for the 3, 4 and 6 times tables

Example

You should know your 2, 5 and 10 times tables.

×	1	2	3	4	5	6	7	8	9	10
1	1	2	3	4	5	6	7	8	9	10
2	2	4	6	8	10	12	14	16	18	20
3	3	6	9	12	15	18	21	24	27	30
4	4	8	12	16	20	24	28	32	36	40
5	5	10	15	20	25	30	35	40	45	50
6	6	12	18	24	30	36	42	48	54	60

You know that
2 × 8 = 16

4 is double 2.
So 4 × 8 is
double 2 × 8

Learn the 3 times table. You can then use it to
work out the 6 times table.

Key facts

Numbers that are multiples of 2 are called **even numbers**.
The order does not matter when you multiply two numbers:
3 × 4 = 4 × 3 = 12

Language

Multiplication fact: For each multiplication fact like 5 × 6 = 30 there is
another multiplication fact: 6 × 5 = 30.
30 is the **product** of 5 and 6.
30 is a **multiple** of 5 and 30 is a **multiple** of 6.

On track

1 Which of these numbers are even? Circle them.

113   216   317   419   522   625   726   834   950

2 Which of these numbers are multiples of 5? Circle them.

110   213   315   419   520   623   725   829   930

3 Which of these numbers are multiples of by 10? Circle them.

101   240   385   465   570   610   773   820   999

4 For questions 1, 2 and 3, how did you know which ones to circle?

Aiming higher

1 Try to do these from memory.

   a) 2×6 **b)** 7×2 **c)** 2×8 **d)** 5×2 **e)** 2×9 **f)** 4×2

2 Try to do these from memory.

   a) 5×6 **b)** 7×5 **c)** 5×8 **d)** 5×5 **e)** 5×9 **f)** 4×5

3 Complete these number statements using doubling:

   a) $2 \times 4 = 8$ so $4 \times 4 = ?$ **b)** $2 \times 5 = 10$ so $4 \times 5 = ?$

4 Complete these number statements using doubling:

   a) $3 \times 4 = 12$ so $6 \times 4 = ?$ **b)** $3 \times 5 = 15$ so $6 \times 5 = ?$

5 For $3 \times 6 = 18$, write another multiplication fact.

Using and applying

1 I am thinking of an even number between 10 and 20. It is a multiple of 3 and 4. What is it?

2 I am thinking of an even number between 20 and 30. It is a multiple of 3 and 4. What is it?

3 I am thinking of an even number between 1 and 100. It is a multiple of 4 and 6 and 10. What is it?

4 Make up a number puzzle of your own and test it on a friend.

8 Division facts

What do you need to know?

- Multiplication and division facts for the 2, 5 and 10 times tables
- How to recognise multiples of 2, 5 or 10

What will you learn?

- How to remember all the division facts for the 3, 4 and 6 times tables

Example

You should know your 2, 5 and 10 times tables.

- For each multiplication fact, there are two division facts.

	1	2	3	4	5	6	7	8	9	10
2	2	4	6	8	10	12	14	16	18	20
5	5	10	15	20	25	30	35	40	45	50
10	10	20	30	40	50	60	70	80	90	100

$5 \times 6 = 30$
$30 \div 5 = 6$
$30 \div 6 = 5$

- The same is true for the 3, 4 and 6 times tables.

	1	2	3	4	5	6	7	8	9	10
3	3	6	9	12	15	18	21	24	27	30
4	4	8	12	16	20	24	28	32	36	40
6	6	12	18	24	30	36	42	48	54	60

$4 \times 6 = 24$
$24 \div 4 = 6$
$24 \div 6 = 4$

Key facts

Numbers that are multiples of 2 are called **even numbers**.

Language

For each **multiplication fact** such as $5 \times 6 = 30$, two **matching division facts** link the three numbers: $30 \div 5 = 6$ and $30 \div 6 = 5$.

On track

1 For each of these divisions, write a multiplication fact that will help you to find the answer.

    a) 12 ÷ 6   **b)** 14 ÷ 2

    c) 16 ÷ 8   **d)** 10 ÷ 2

    e) 18 ÷ 9   **f)** 8 ÷ 2

2 Do these divisions. What multiplication fact did you use to help you?

    a) 30 ÷ 6   **b)** 15 ÷ 5

    c) 40 ÷ 8   **d)** 20 ÷ 4

    e) 25 ÷ 5   **f)** 35 ÷ 7

    g) 30 ÷ 10   **h)** 50 ÷ 5

    i) 40 ÷ 4

Aiming higher

1 For each of these divisions, write a multiplication fact that will help you to find the answer.

    a) 30 ÷ 3   **b)** 12 ÷ 4   **c)** 18 ÷ 6

    d) 27 ÷ 9   **e)** 15 ÷ 3   **f)** 9 ÷ 3

2 Do these divisions. What multiplication fact did you use to help you?

    a) 24 ÷ 6   **b)** 16 ÷ 4   **c)** 32 ÷ 8   **d)** 21 ÷ 3

    e) 12 ÷ 3   **f)** 28 ÷ 7   **g)** 36 ÷ 6   **h)** 24 ÷ 4

    i) 48 ÷ 8   **j)** 18 ÷ 3   **k)** 54 ÷ 6   **l)** 30 ÷ 5

3 For each of these division facts, write another division fact and two matching multiplication facts.

    a) 40 ÷ 5 = 8   **b)** 42 ÷ 7 = 6   **c)** 45 ÷ 5 = 9

Using and applying

1 I am thinking of a number between 10 and 25. It is a multiple of 3 and a multiple of 4. What numbers could it be?

2 I am thinking of an even number between 25 and 50. It is a multiple of 3 and a multiple of 4. What numbers could it be?

3 I am thinking of an odd number between 1 and 50. It is a multiple of 3 and a multiple of 5. What numbers could it be?

4 Make up a number puzzle of your own and test it on a friend.

9 Double checking

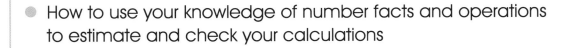

What do you need to know?

- How to use your knowledge of number facts and operations to estimate and check your calculations

What will you learn?

- How to use your knowledge of number operations and corresponding inverses to estimate and check your calculations
- How to use your knowledge of doubling and halving to estimate and check your calculations

Example

- Katie has worked out that 75 – 27 = 48. How can she check her answer?

  *Addition is the inverse of subtraction. So, to double-check her answer, Katie can write a **matching addition sum** and work that out.*

  27 + 48 = 20 + 7 + 40 + 8 = 60 + 15 = 75 ✓

- Katie could also **estimate** the answer.
  75 is close to 80 and 27 is close to 30.

  80 – 30 = 50 which is close to her answer of 48.

Key facts

Addition and subtraction are **inverse operations**.

Language

Inverse operations: 'undo' each other.
For each **addition** fact, such as, 23 + 41 = 64, there is another addition
  fact: 41 + 23 = 64, and two **subtraction facts** that match: 64 – 23 =
  41 and 64 – 41 = 23.
An **estimate:** this is close to the correct answer. It is not a guess.
Approximating numbers makes arithmetic easy to do mentally.

On track

1. For each of these addition facts, write another addition fact and two corresponding subtraction facts.
   a) $23 + 54 = 77$
   b) $11 + 68 = 79$

2. How can knowing that $30 + 50 = 80$ help you to estimate the answers to these sums?
   a) $84 + 27$
   b) $81 - 53$

3. What is $28 \div 4$? Write a multiplication fact that you can use to check your answer.

4. Half of 32 is 16. Write a sentence that uses the word 'double' and the numbers 16 and 32.

Aiming higher

1. Use estimation to find an approximate answer to these sums. Then calculate your answer. How close was your estimate?
   a) $82 - 11$
   b) $28 + 39$
   c) $53 - 29$
   d) $31 + 18$

2. Use estimation to check these answers. For the ones that are wrong, calculate the correct answer.
   a) $35 - 16 = 9$
   b) $28 + 29 = 47$
   c) $53 - 27 = 26$
   d) $31 - 18 = 49$

Using and applying

1. Katie works out that $82\,cm - 38\,cm = 56\,cm$. Estimate the answer to check whether her answer is right. Recalculate the sum as a second check.

2. Paul has £20. He buys toys costing £6.75 and £2.85. Will his change be closer to £12, £11 or £10? Estimate the answer and then do the calculation.

3. Use inverse operations to solve this problem. *I think of a number, double it and then take away 3. My answer is 7. What was my number?*

4. Make up a number puzzle of your own and test it on a friend.

10 Mental addition and subtraction

What do you need to know?

- How to add a 1-digit number to a 2-digit number
- How to add a multiple of 10 to a 2-digit number
- How to subtract a 1-digit number from a 2-digit number
- How to subtract a multiple of 10 from a 2-digit number

What will you learn?

- How to add 2-digit numbers in your head
- How to subtract 2-digit numbers in your head

Example

- How could you add 15 and 24 in your head?

You know how to add a 1-digit number, like 4, to 15.

*You know how to add a multiple of 10, like 20, to 15. So, **partition** 24 into 20 + 4 and do the mental calculation in two steps:*

↗Add 4↘   ↗Add 20↘

15 + 4 = 19   19 + 20 = 39

Key facts

The order you add the 4 and the 20 does not matter.

```
         Add 20              Add 4
15           →  35      →        39
```

You could also add 15 to 24, in any order.
The answer is the same.

24 + 5 = 29   29 + 10 = 39   and   24 + 10 = 34   34 + 5 = 39

TIP! Decide on an order that suits you and then stick to it.

Language

Partitioning: breaking down a number into smaller numbers to help you to do mental arithmetic, e.g. 15 = 10 + 5 and 24 = 20 + 4.

On track

1 Do these sums in your head.
a) 12 + 7   b) 25 + 3
c) 31 + 5   d) 44 + 8
e) 53 + 9   f) 65 + 6

2 Do these sums in your head.
a) 12 + 70   b) 25 + 30
c) 31 + 50   d) 44 + 80
e) 53 + 90   f) 65 + 60

Aiming higher

1 Do these sums in your head. First add the units digit of the second number. Then add the tens digit.
a) 42 + 17   b) 45 + 23   c) 51 + 35
d) 44 + 48   e) 13 + 39   f) 25 + 76

2 Do these sums in your head. First add the tens digit of the second number. Then add the units digit.
a) 52 + 17   b) 35 + 23   c) 61 + 35
d) 24 + 48   e) 23 + 59   f) 15 + 19

3 Do these sums in your head. Notice you are adding a smaller number to a bigger number.
a) 19 + 17   b) 25 + 23   c) 36 + 35
d) 49 + 48   e) 53 + 39   f) 65 + 16

4 Do these sums in your head. Notice you are adding a bigger number to a smaller number.
a) 12 + 17   b) 15 + 23   c) 21 + 35
d) 24 + 48   e) 33 + 39   f) 15 + 16

Using and applying

1 Which method of adding do you prefer? Adding tens first and then units? Or adding units first and then tens? Are you more accurate using one method than the other?

2 Is it easier to add a smaller number to a bigger number, such as 19 to 45? Or, is it easier to add a bigger number to a smaller number, such as 45 to 19? Which method suits you better?

11 Written addition

What do you need to know?

● The addition facts for pairs of 1-digit numbers

What will you learn?

● How to write addition sums of 2-digit and 3-digit numbers

Example

Here are some ways of writing addition sums.

● Partitioning works like this:

$$35 + 42 \qquad\qquad\qquad 35 + 49$$

$$30 + 5 \quad + \quad 40 + 2 \qquad\qquad 30 + 5 \quad + \quad 40 + 9$$

$$30 + 40 \quad + \quad 5 + 2 \qquad\qquad 30 + 40 \quad + \quad 5 + 9$$

$$70 \quad + \quad 7 \qquad\qquad\qquad 70 \quad + \quad 14$$

$$77 \qquad\qquad\qquad\qquad 84$$

● Setting out your sum in columns looks like this:

Add with no carry   *Add with carry*

```
H T U                                    H T U              H T U
1 3 5        Be sure to line             1 3 5              5 3 5
2 4 2 +      up the columns              2 4 9 +            1 4 9 +
             carefully.                                         1
─────                                    ─────              ─────
3 7 7                                    3 8 4              6 8 4
                                            1
```

A carry digit can be written in the T column below this line ... or in the T column above this line.

Key facts

Each column is worth ten times more than the one to its right.

Language

H stands for 'hundreds'.   **T** stands for 'tens'.   **U** stands for 'units'.

On track

1 Do these sums in your head.

a) 7 + 9   **b)** 8 + 6   **c)** 6 + 7   **d)** 8 + 9

e) 7 + 7   **f)** 6 + 9   **g)** 7 + 8   **h)** 8 + 8

2 Try to do these sums in your head, but if you need to, show your working.

a) 17 + 9   **b)** 28 + 6   **c)** 36 + 7   **d)** 48 + 9

e) 17 + 27   **f)** 16 + 19   **g)** 37 + 18   **h)** 48 + 15

3 Do these sums, showing your working.

a) 34 + 25   **b)** 72 + 17   **c)** 241 + 114   **d)** 355 + 132

 Aiming higher

1 Try to do these sums in your head but, if you need to, show your working.

a) 27 + 19   **b)** 38 + 16   **c)** 46 + 27   **d)** 58 + 19

e) 37 + 25   **f)** 66 + 17   **g)** 57 + 28   **h)** 78 + 15

2 Do these sums, showing your working.

a) 38 + 25   **b)** 76 + 17   **c)** 247 + 414   **d)** 559 + 532

 Using and applying

1 The teacher has marked Katie's sums as wrong. Check the workings to find Katie's mistakes. Re-do the sums, correctly.

a) 145
      238 +
      ‾‾‾‾‾
      473 ✗

b) 782
      127 +
      ‾‾‾‾‾
      809 ✗

c) 396
      144 +
      ‾‾‾‾‾
      252 ✗

2 Compare your answers to all the sums on this page with someone else's. If you disagree on an answer, check both sums and see where the mistake has been made.

12 Written subtraction

What do you need to know?

- The subtraction facts for pairs of numbers with totals up to 20

What will you learn?

- How to write subtraction sums of 2-digit and 3-digit numbers

Example

Here are some ways of writing subtraction sums.

- Partitioning works like this:

72 – 31	72 – 39
70 + 2 – 30 – 1	60 + 12 – 30 – 9
70 – 30 + 2 – 1	60 – 30 + 12 – 9
40 + 1	30 + 3
41	33

Remember to subtract the Ts and the Us

Partition so that the 9 can be subtracted from 12.

- Setting out your sum in columns looks like this:

$$
\begin{array}{r}
7\,2 \\
3\,1\, - \\
\hline
4\,1 \\
\end{array}
\qquad
\begin{array}{r}
^{60}\!\!\!\not{7}\,^{1}\!2 \\
3\,9\, \\
\hline
3\,3 \\
\end{array}
$$

9 from 2 won't go so borrow a 10 from the 70

Make sure you line up the columns.

Key facts

Each column is worth ten times more than the one to its right.

Language

H stands for 'hundreds'.   **T** stands for 'tens'.   **U** stands for 'units'.

On track

1   Do these sums in your head.
   a) 17 – 9   b) 18 – 6   c) 16 – 7   d) 18 – 9
   e) 17 – 7   f) 16 – 9   g) 17 – 8   h) 18 – 8

2   Try to do these sums in your head, but if you need to, show
your workings.
   a) 27 – 9   b) 38 – 6   c) 46 – 7   d) 58 – 9
   e) 67 – 7   f) 76 – 9   g) 87 – 8   h) 98 – 8

3   Do these sums, showing your workings.
   a) 234 – 122   b) 472 – 171   c) 241 – 140   d) 355 – 232

Aiming higher

1   Try to do these sums in your head but, if you need to, show
your workings.
   a) 37 – 19   b) 48 – 16   c) 66 – 17   d) 58 – 19
   e) 77 – 17   f) 86 – 19   g) 97 – 18   h) 89 – 18

2   Do these sums, showing your workings.
   a) 538 – 329   b) 276 – 117   c) 347 – 149   d) 591 – 328

Using and applying

1   The teacher has marked Katie's sums as wrong. Check the workings to
find Katie's mistakes. Re-do the sums, correctly.

   a)   645   b)   782   c)   396
        238 –   127 –   144 –
       ─────   ─────   ─────
        413 ✗   665 ✗   540 ✗
       ─────   ─────   ─────

2   Compare your answers to the sums on this page with someone else in
your class. If you disagree on an answer, check both sums and see where
the mistake has been made.

13 Multiplying by 10 or 100

What do you need to know?

- How to use repeated addition for multiplication

What will you learn?

- How to multiply 1-digit and 2-digit numbers by 10
- How to multiply 1-digit and 2-digit numbers by 100

Example

- To work out 4 times 10, you could add ten 4s together.

$4 \times 10 = 4 + 4 + 4 + 4 + 4 + 4 + 4 + 4 + 4 + 4 = 40$

- To work out 60 times 10, you could add ten 60s together.

$60 \times 10 = 60 + 60 + 60 + 60 + 60 + 60 + 60 + 60 + 60 + 60 = 600$

> Check that you agree with these totals.

- To work out 10 times 64, use partitioning.

$10 \times 64 = 10 \times (60 + 4) = 10 \times 60 + 10 \times 4$

$= 600 + 40 = 640$

Notice that the 6 that was in the T column appears in the H column.
The 4 that was in the U column appears in the T column.

Key facts

Look what happens when you multiply the single digit 4 by 10.

U   TU   TU
$4 \times 10 = 40$

The 4 appears in the Tens column and a 0 stands as a placeholder in the Units column.
When you multiply the 2-digit number 60 by 10, the 6 appears in the Hundreds column and 0s appear in the Tens and Units columns as placeholders.

TU   TU   HTU
$60 \times 10 = 600$

When you multiply, the digits do not just 'move'. They increase in value tenfold. What was worth 6 tens becomes worth 6 hundreds. They appear in the next column to the left of wherever they were before.

Language

Placeholder: a zero which shows there are none for that column in the number.

 **On track**

1. **a)** Do these additions: $3 + 3 + 3 + 3$ and $4 + 4 + 4$

   b) Do these multiplications: 3×4 and 4×3

   c) Count these squares.

   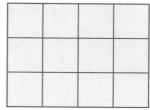

   d) Compare your answers to parts **a)** and **b)**. Do they agree? How could they be used to help you to answer part **c)**?

2. Do these additions.

   a) $6 + 6 + 6 + 6 + 6 + 6 + 6 + 6 + 6 + 6$

   b) $8 + 8 + 8 + 8 + 8 + 8 + 8 + 8 + 8 + 8$

   c) $30 + 30 + 30 + 30 + 30 + 30 + 30 + 30 + 30 + 30$

   d) $50 + 50 + 50 + 50 + 50 + 50 + 50 + 50 + 50 + 50$

   e) $70 + 70 + 70 + 70 + 70 + 70 + 70 + 70 + 70 + 70$

3. Do these multiplications.

   a) 6×10 **b)** 8×10 **d)** 30×10

   e) 50×10 **f)** 70×10

4. Compare your answers to questions 2 and 3. Do they agree?

 Aiming higher

1. Do these multiplications.

   a) 2×100 **b)** 6×100 **c)** 8×100

2. Use partitioning to do these multiplications.

   a) 32×10 **b)** 56×10 **c)** 78×10

   d) 23×100 **e)** 65×100 **f)** 87×100

 Using and applying

1. What are the next two numbers in this sequence? 4, 40, 400, ... , ...

2. I am thinking of a number. I multiple it by 10 and add 30. I get 430. What was my number?

14 Written multiplication

What do you need to know?

- How to show repeated addition and arrays as multiplication

What will you learn?

- How to write multiplication calculations

Example

- You can show 13 × 4 as an array and count the squares.

1	2	3	4	5	6	7	8	9	10	11	12	13
14	15	16	17	18	19	20	21	22	23	24	25	26
27	28	29	30	31	32	33	34	35	36	37	38	39
40	41	42	43	44	45	46	47	48	49	50	51	52

Partitioning can also be used to work out a multiplication such as 13 × 4.

$$13 = 10 + 3 \qquad\qquad 13 \times 4 = 10 \times 4 + 3 \times 4$$
$$= 40 + 12 = 52$$

Multiplication calculations can also be set out in a grid.

×	10	3	13 × 4
4	40	12	52

4 × 10 = 40	4 × 3 = 12	40 + 12 = 52

Key facts

When you multiply one number by another you must multiply each digit.
$$13 \times 4 = (10 \times 4) + (3 \times 4)$$
Be careful not to lose the placeholder zeroes.

The zero is a placeholder.

Language

Product: the result of multiplying two numbers together, e.g. 52 is the product of 13 and 4.

RISING STARS

Mathematics Study Guide: Year 3

Answer Booklet

Unit 1 Numbers up to 1000
On track
1 a) 245   **b)** 608
2 10, 12, 13, 15, 16, 17
3 15, 37, 51, 56, 65, 73, 89, 98

Aiming higher
1 a) 4   **b)** 3   **c)** 6
2 a) 101   **b)** 99   **c)** 100
3 751, 750, 715, 705, 575, 570, 515, 157

Using and applying
1 28
2 9°C

Unit 2 Partitioning 3-digit numbers
On track
1 a) 40   **b)** 4   **c)** 400   **d)** 40
2 a) 30 + 6   **b)** 70 + 3
   c) 40 + 1   **d)** 80 + 2
3 a) 100 + 30 + 6   **b)** 200 + 90 + 1
   c) 300 + 70 + 2   **d)** 400 + 20 + 5

Aiming higher
1 a) 200 + 40 + 5 = 200 + 30 + 15
   b) 300 + 80 + 7 = 300 + 70 + 17
   c) 400 + 30 + 9 = 400 + 20 + 19
   d) 500 + 20 + 1 = 500 + 10 + 11
2 a) 100, 100, 10, 1, 10, 1
   b) 264, 462

Using and applying
1 67, 77, 87, 97, 107
2 554, 454, 354, 254, 154, 54

Unit 3 Rounding and estimating
On track
1 a) 60   **b)** 80   **c)** 30   **d)** 40
   e) 40   **f)** 50
2 a) 520   **b)** 860   **c)** 230   **d)** 670
   e) 330   **f)** 750
3 a) 500   **b)** 900   **c)** 200   **d)** 700
   e) 300   **f)** 700
4 a) 500   **b)** 900   **c)** 200   **d)** 700
   e) 300   **f)** 800

Aiming higher
1 a) 30 + 30 = 60
   b) 50 + 50 = 100
   c) 80 + 20 = 100
   d) 60 + 80 = 140
2 a) 400 + 300 = 700
   b) 100 + 900 = 1000
   c) 300 + 700 = 1000
   d) 600 + 300 = 900
3 59, 102, 101, 136; 765, 997, 1001, 836

Using and applying
1 8 x 30 = 240
2 10 x 20 = 200
3 31 + 28 + 31 + 30 + 31 + 30 + 31
   + 31 = 243;
   12 x 19 = 228

Unit 4 Proper fractions
On track
1 $\frac{1}{7}$ and $\frac{5}{9}$
2 $\frac{1}{4}$, $\frac{1}{2}$, $\frac{3}{4}$
3 $\frac{6}{8} = \frac{3}{4}$
4

Aiming higher
1 $\frac{1}{3}$, $\frac{2}{6}$, $\frac{3}{12}$
2 $\frac{1}{3} = \frac{2}{6} = \frac{3}{9}$
3 $\frac{2}{3}$, $\frac{4}{6}$, $\frac{3}{9}$
4 $\frac{2}{3} = \frac{4}{6} = \frac{6}{9}$

Using and applying
1
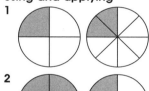
2
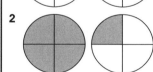

Unit 5 Addition facts
On track
1 a) 9   **b)** 7   **c)** 5   **d)** 3
2

	11	12	13	14	15	16	17	18	19	20
11	22	23	24	25	26	27	28	29	30	31
12	23	24	25	26	27	28	29	30	31	32
13	24	25	26	27	28	29	30	31	32	33
14	25	26	27	28	29	30	31	32	33	34
15	26	27	28	29	30	31	32	33	34	35
16	27	28	29	30	31	32	33	34	35	36
17	28	29	30	31	32	33	34	35	36	37
18	29	30	31	32	33	34	35	36	37	38
19	30	31	32	33	34	35	36	37	38	39
20	31	32	33	34	35	36	37	38	39	40

Aiming higher
1 a) 59   **b)** 27   **c)** 65   **d)** 33
2 a) 80   **b)** 80   **c)** 50   **d)** 50
   e) 70   **f)** 40

Using and applying
1 6 + 5 = 11; 2 + 7 = 9; 6 + 2 = 8;
   5 + 7 = 12; 11 + 9 = 8 + 12 = 20

Unit 6 Subtraction facts
On track
1 a) 6, 6 + 4 = 10   **b)** 8, 8 + 2 = 10
   c) 4, 4 + 6 = 10   **d)** 2, 2 + 8 = 10
2 a) 60, 10 + 60 = 70
   b) 70, 20 + 70 = 90
   c) 60, 20 + 60 = 80
   d) 50, 10 + 50 = 60

Aiming higher
1 a) 16, 16 + 4 = 20
   b) 18, 18 + 2 = 20
   c) 14, 14 + 6 = 20
   d) 12, 12 + 8 = 20
2 e.g. 10 + 5 = 15, 15 − 5 = 10,
   15 − 10 = 5
3 a) e.g. 7 + 9 = 16, 16 − 7 = 9,
   16 − 9 = 7
   b) e.g. 12 + 5 = 17, 17 − 12 = 5,
   17 − 5 = 12
   c) e.g. 14 + 4 = 18, 18 − 4 = 14,
   18 − 14 = 4
   d) e.g. 11 + 8 = 19, 19 − 8 = 11,
   19 − 11 = 8
4 a) 57, 57 + 43 = 100
   b) 73, 73 + 27 = 100
   c) 38, 38 + 62 = 100
   d) 11, 11 + 89 = 100

Using and applying
1 16 + 28 = 44; 16 + 25 = 41;
   25 + 31 = 56; 44 + 56 = 100;
   41 + 59 = 100
2 Any four numbers that sum to 100.

Unit 7 Multiplication facts
On track
1 216, 522, 726, 834, 950
2 110, 315, 520, 725, 930
3 240, 570, 610, 820
4 Even numbers end in 0, 2, 4, 6 or 8;
   Multiples of 5 end in 5 or 0.
   Multiples of 10 end in 0.

Aiming higher
1 a) 12   **b)** 14   **c)** 16   **d)** 10
   e) 18   **f)** 8
2 a) 30   **b)** 35   **c)** 40   **d)** 25
   e) 45   **f)** 20
3 a) 16   **b)** 20
4 a) 24   **b)** 30
5 6 x 3 = 18

Using and applying
1 12
2 24
3 60

Unit 8 Division facts
On track
1 a) 2, 2 x 6 = 12   **b)** 7, 2 x 7 = 14
   c) 2, 2 x 8 = 16   **d)** 5, 2 x 5 = 10
   e) 2, 2 x 9 = 18   **f)** 4, 2 x 4 = 8
2 a) 5, 5 x 6 = 30   **b)** 3, 3 x 5 = 15
   c) 5, 5 x 8 = 40   **d)** 5, 4 x 5 = 20
   e) 5, 5 x 5 = 25   **f)** 5, 5 x 7 = 35
   g) 3, 3 x 10 = 30   **h)** 5, 5 x 10 = 50
   i) 4, 4 x 10 = 40

Aiming higher
1 a) 10, 3 x 10 = 30   **b)** 3, 3 x 4 = 12
   c) 3, 3 x 6 = 18   **d)** 3, 3 x 9 = 27

e) 5, 3 x 5 = 15 **f)** 3, 3 x 3 = 9
2 a) 4, 4 x 6 = 24 **b)** 4, 4 x 4 = 16
c) 4, 4 x 8 = 32 **d)** 7, 3 x 7 = 21
e) 4, 3 x 4 = 12 **f)** 4, 4 x 7 = 28
g) 6, 6 x 6 = 36 **h)** 6, 4 x 6 = 24
i) 6, 6 x 8 = 48 **j)** 6, 3 x 6 = 18
k) 9, 9 x 6 = 54 **l)** 6, 5 x 6 = 30
3 a) 40 ÷ 8 = 5, 5 x 8 = 40, 8 x 5 = 40
b) 42 ÷ 6 = 7, 6 x 7 = 42, 7 x 6 = 42
c) 45 ÷ 9 = 5, 5 x 9 = 45, 9 x 5 = 45

Using and applying
1 12, 24
2 36, 48
3 15, 45

Unit 9 Double checking
On track
1 a) 54 + 23 = 77, 77 – 23 = 54,
77 – 54 = 23
b) 68 + 11 = 79, 79 – 11 = 68,
79 – 11 = 68
2 a) 80 – 30 = 50, so
84 – 27 = 50 + 4 + 3 = 57
b) 80 – 50 = 30, so
81 – 53 = 50 + 1 – 3 = 48
3 28 ÷ 4 = 7, 4 x 7 = 28
4 32 is double 16.

Aiming higher
1 a) 80 – 10 = 70;
82 – 11 = 70 + 2 – 1 = 71
b) 30 + 40 = 70;
28 + 39 = 70 – 2 – 1 = 67
c) 50 – 30 = 20;
53 – 29 = 20 + 3 + 1 = 24
d) 30 + 20 = 50;
31 + 18 = 50 + 1 – 2 = 49
2 a) 35 – 16 = 19 **b)** 28 + 29 = 57
c) 31 – 18 = 13

Using and applying
1 80 – 40 – 40; wrong. 82 – 38 = 44
2 £7 + £3 = £10; change £10.
£6.75 + £2.85 = £9.60.
Change £10.40
3 2x – 3 = 7; x = 5

Unit 10 Mental addition and subtraction
On track
1 a) 19 **b)** 28 **c)** 36 **d)** 52;
e) 62 **f)** 71
2 a) 82 **b)** 55 **c)** 81 **d)** 124
e) 143 **f)** 125

Aiming higher
1 a) 59 **b)** 68 **c)** 86 **d)** 92
e) 52 **f)** 101
2 a) 69 **b)** 58 **c)** 96 **d)** 72
e) 82 **f)** 34
3 a) 36 **b)** 48 **c)** 71 **d)** 97
e) 92 **f)** 81
4 a) 29 **b)** 38 **c)** 56 **d)** 72
e) 72 **f)** 31

Unit 11 Written addition
On track
1 a) 16 **b)** 14 **c)** 13 **d)** 17
e) 14 **f)** 15 **g)** 15 **h)** 16
2 a) 26 **b)** 34 **c)** 43 **d)** 57
e) 44 **f)** 35 **g)** 55 **h)** 63
3 a) 59 **b)** 89 **c)** 355 **d)** 487

Aiming higher
1 a) 46 **b)** 54 **c)** 73 **d)** 77
e) 62 **f)** 83 **g)** 85 **h)** 93
2 a) 63 **b)** 93 **c)** 661 **d)** 1091

Using and applying
1 a) 383 **b)** 909 **c)** 540

Unit 12 Written subtraction
On track
1 a) 8 **b)** 12 **c)** 9 **d)** 9
e) 10 **f)** 7 **g)** 9 **h)** 10
2 a) 18 **b)** 32 **c)** 39 **d)** 49
e) 60 **f)** 67 **g)** 79 **h)** 90
3 a) 112 **b)** 301 **c)** 101 **d)** 123

Aiming higher
1 a) 18 **b)** 32 **c)** 49 **d)** 39
e) 60 **f)** 67 **g)** 79 **h)** 71
2 a) 209 **b)** 159 **c)** 198 **d)** 263

Using and applying
1 a) 383 **b)** 909 **c)** 540

Unit 13 Multiplying by 10 or 100
On track
1 a) 12, 12 **b)** 12, 12 **c)** 12
d) 3 + 3 + 3 + 3 = 4 x 3 = 12;
4 + 4 + 4 = 3 x 4 = 12
2 a) 60 **b)** 80 **c)** 300 **d)** 500
e) 700
3 a) 60 **b)** 80 **c)** 300 **d)** 500
e) 700
4 Yes

Aiming higher
1 a) 200 **b)** 600 **c)** 800
2 a) 320 **b)** 560 **c)** 780 **d)** 2300
e) 6500 **f)** 8700

Using and applying
1 4000, 40,000
2 10x + 30 = 430; x = 40

Unit 14 Written multiplication
On track
1 a) 32 **b)** 18 **c)** 45 **d)** 48
e) 36 **f)** 42
2 a) 36 **b)** 78 **c)** 156 **d)** 230
e) 288 **f)** 235

Aiming higher
1 a) 324 **b)** 310 **c)** 558 **d)** 448
e) 680 **f)** 684
2 a) 72 **b)** 130 **c)** 234 **d)** 322
e) 416 **f)** 549

Using and applying
1 15 cm + 15 cm + 15 cm + 15 cm =
4 x 15 cm = 60 m
2 8 x 12 cm = 96 cm

Unit 15 Written division
On track
1 a) 2 **b)** 5 **c)** 7 **d)** 9
e) 11 **f)** 12

Aiming higher
1 a) 6r2 **b)** 4r4 **c)** 5r3 **d)** 9r2
e) 8r2 **f)** 8r4
2 a) 7 **b)** 8 **c)** 7 **d)** 11
e) 8 **f)** 5
3 a) 12 **b)** 9 **c)** 9 **d)** 6
e) 16 **f)** 6

Using and applying
1 27 ÷ 6 = 4r3; 4 groups, 3 left over
2 4 rows (round down)
3 5 boxes (round up)

Unit 16 Division as the inverse of multiplication
On track
1 a) 9 x 6 = 54 **b)** 6 x 7 = 42
c) 4 x 5 = 20 **d)** 8 x 4 = 32
2 a) 54 ÷ 6 = 9 or 54 ÷ 9 = 6
b) 42 ÷ 7 = 6 or 42 ÷ 6 = 7
c) 20 ÷ 5 = 4 or 20 ÷ 4 = 5
d) 32 ÷ 4 = 8 or 32 ÷ 8 = 4

Aiming higher
1 a) 9 x 7 = 63, 63 ÷ 7 = 9, 63 ÷ 9 = 7
b) 6 x 8 = 48, 48 ÷ 6 = 8, 48 ÷ 8 = 6
c) 4 x 7 = 28, 28 ÷ 7 = 4, 28 ÷ 4 = 7
d) 8 x 5 = 40, 40 ÷ 8 = 5, 40 ÷ 5 = 8
2 a) 8 x 9 = 9 x 8 = 72, 72 ÷ 8 = 9,
72 ÷ 9 = 8
b) 7 x 8 = 8 x 7 = 56, 56 ÷ 7 = 8,
56 ÷ 8 = 7
c) 4 x 6 = 6 x 4 = 24, 24 ÷ 6 = 4,
24 ÷ 4 = 6
d) 5 x 7 = 7 x 5 = 35, 35 ÷ 5 = 7,
35 ÷ 7 = 5

Using and applying
1 30 ÷ 5 = 6; 5 x 6 = 30
2 18 x 5 ÷ 5 = 18; 5 ÷ 5 = 1

Unit 17 Unit fractions
On track
1 a) 9 **b)** 7 **c)** 6 **d)** 8
e) 6 **f)** 5 **g)** 3 **h)** 6
i) 7
2 a) 10 cm **b)** 8 ml **c)** 5 g **d)** 7 ml
e) 9 g **f)** 4 cm **g)** 4 g **h)** 5 cm

Aiming higher
1 a) 2 m **b)** 4 litres **c)** 3 kg **d)** 3 kg
e) 6 km **f)** 9 ml
2 a) 500 cm **b)** 500 ml
c) 500 g **d)** 250 g
e) 25 cm **f)** 250 ml
3 9 p

Using and applying
1 15 mm
2 6 days; 20 ml; $\frac{120}{6}$
3 20 g; $\frac{1}{5}$

Unit 18 2-D and 3-D shapes
On track
1 a) rectangle **b)** right-angled triangle
c) circle **d)** isosceles triangle
e) hexagon **f)** square
g) pentagon **h)** trapezium **i)** kite
j) rhombus/diamond **k)** oval/ellipse
2 a) sphere **b)** cube **c)** cuboid
d) cone **e)** triangular prism
f) cylinder **g)** hexagonal prism
h) pentagonal prism

Aiming higher
1 square + double square (= rectangle);
quarter and semi circles; arrowhead
and triangle (half of arrowhead);
hexagon and trapezium
2 a) cuboid (no triangular faces)
b) cylinder (curved surface)
c) cube (no circular faces)

Unit 19 Reflections in a mirror
On track
1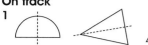

2 Vertical line of symmetry: A, H, I, M, T, U, V, W, Y.
Horizontal: B, C, D, E, H, I.
Vert + horizontal: O, X, H, I.

3

Aiming higher
1 a) b) c)

2 WAY OUT
3 *Own answer*

Unit 20 Position, direction and movement
On track
1 a) E   b) S   c) E   d) W
2 a) N   b) E   c) N   d) S

Aiming higher
1 a) X in 3D   b) Y in 5F   c) Z in 1A
2 *Own answer*

Unit 21 Right angles in 2D shapes
On track
1 acute: b, d, f; obtuse: a, e; right c
2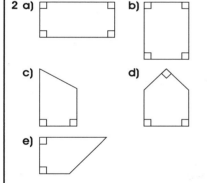

Aiming higher
1 a)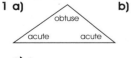
obtuse
acute   acute

b)
acute
right   acute

c)
acute
obtuse   acute

2 a) regular pentagon   b) irregular pentagon

c) irregular pentagon

3 a) regular hexagon   b) irregular hexagon

c) irregular hexagon

Using and applying
1 yes
2 no

Unit 22 Units of length, weight and capacity
On track
1 a) 5 m   b) 6 km   c) 6 kg   d) 5 litres
2 a) 400 cm   b) 2000 m
   c) 3000 g   d) 7000 ml
3 b) 500 ml c) 50 cm d) 500 m e) 250 g
   f) 250 ml

Aiming higher
1 a) kilograms   b) grams
   c) milligrams
2 a) mm   b) cm   c) km   d) m

Using and applying
1 25 cm
2 500 g
3 3 cartons

Unit 23 Reading scales
On track
1 1.8 cm, 3.4 cm, 8.1 cm, 11.6 cm
2 15 cm; 12 cm; 8.5 cm; 5 cm (all to nearest 0.5 cm)

Aiming higher
1 centimetres or millimetres;
   height = 29.6 cm; width = 210 mm
2 grams

Unit 24 Intervals of time
On track
1 60 seconds; 60 minutes; 24 hours
2 5 past, 10 past, 20 past, 25 past, 25 to, 20 to, 10 to, 5 to
3 a) 1:10   b) 5:40   c) 9:20   d) 10:50

Aiming higher
1 a) b)
c) d)

2 12:00; 12:00 (or 0:00)

Using and applying
1 20 minutes
2 40 minutes

Unit 25 Tally charts and frequency tables
On track
1

Value	Tally	Frequency
1	卌	5
2	卌 I	6
3	卌	5
4	IIII	4
5	卌 II	7
6	III	3
		30 ✓

2 Sum of frequency = 40

Using and applying
1 Expect equal frequency for each outcome
2 Lots of times, >60

Unit 26 Using diagrams to sort data
On track
1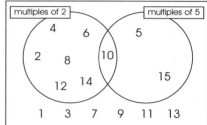

2

Numbers 1 to 20	Multiples of 2	Not multiples of 2
Multiples of 5	10, **20**	5, 15
Not multiples of 5	2, 4, 6, 8, 12, 14, **16**, 18	1, 3, 7, 9, 11, 13, **17**, **19**

Aiming higher
1

Numbers 1 to 20	Multiples of 3	Not multiples of 3
Multiples of 5	15	5, 10
Not multiples of 5	3, 6, 9, 12, 18	1, 2, 4, 7, 8, 11, 13, 14, 16, 17, 19, 20

2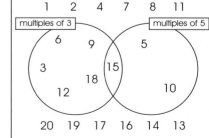

3

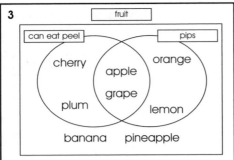

Unit 27 Problem solving
On track
1 50 – 10; 'buys', 'pays with' and 'change'; pence
2 3, 24, 2; 24 ÷ 3 = 8; 8 x 2 = 16

Aiming higher
1 How much I spent altogether; take away from £1
2 10:30, 11:20; 50 minutes; 60 minutes in an hour

Using and applying
1 20 x 5 = 100
2 Use words like 'share'

Unit 28 Using images and diagrams to solve problems
On track
1 6

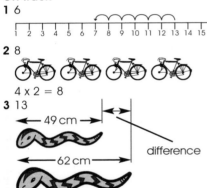

2 8

4 x 2 = 8

3 13

difference

4 D 50 – 30 = 20

Aiming higher
1 £1.20

2 3 grapes per plate

3 $\frac{12}{4}$ = 3

Using and applying
1 For example: 12 cakes are shared between 4 people. How many cakes do they have each?
2 For example, 12 – 5

Unit 29 Following lines of enquiry
On track
1 Use a min/max thermometer to find out how cold it got at night, and check

at noon too; record in degrees; draw a diagram
2 Ask every classmate 'What pets do you have?'; a tally chart, with one row for each category of pet; pie chart or pictogram; largest share, or longest 'bar'.

Aiming higher
1 Use a thermometer, every half hour or so; record on a graph; the temperature rising/falling
2 Write a list of names, and papers required, by address – and also by type of paper (for ordering purposes); write the address on top of each paper

Using and applying
2 Divide a series of consecutive numbers by 3 and record the remainder.

Unit 30 Patterns
On track
1 Multiples of 5 end in 5 or 0
2 80 – 20 = 60; 800 – 600 = 200. The zeroes are place holders
3 Use a set square to check a right angle; use table of results:

1 right angle	2 right angles	More than 2 right angles

Aiming higher
1 857 – 500 = 357
2 Cylinder (curved surface; the other shapes have flat surfaces)

Using and applying
1 83, 78
2 7 + 8 = 15

On track

1 Try to calculate these products in your head.
- **a)** 8 × 4
- **b)** 6 × 3
- **c)** 9 × 5
- **d)** 6 × 8
- **e)** 4 × 9
- **f)** 7 × 6

2 Do these multiplications using partitioning. Show your workings.
- **a)** 18 × 2
- **b)** 26 × 3
- **c)** 39 × 4
- **d)** 46 × 5
- **e)** 48 × 6
- **f)** 47 × 5

Aiming higher

1 Do these multiplications. Show your workings.
- **a)** 81 × 4
- **b)** 62 × 5
- **c)** 93 × 6
- **d)** 64 × 7
- **e)** 85 × 8
- **f)** 76 × 9

2 Do these multiplications. Show your workings.
- **a)** 18 × 4
- **b)** 26 × 5
- **c)** 39 × 6
- **d)** 46 × 7
- **e)** 52 × 8
- **f)** 61 × 9

Using and applying

1 A square lawn has sides 15 metres long.

If you walked around the edge of the lawn, how far would you walk? How did you work this out?

2 Katie wants to wrap this present with ribbon. The present is 12 centimetres long and 12 centimetres wide.

How long a piece of ribbon will Katie need?

How did you work this out?

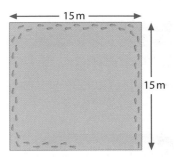

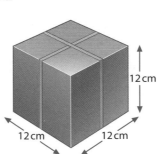

15 Written division

What do you need to know?

- How to show sharing and repeated subtraction as division

What will you learn?

- How to write division calculations and remainders
- When to round up and when to round down

Example

- *Some divisions 'go exactly'.*

  $12 \div 4 = 3$

  Some divisions don't 'go exactly' and there is a remainder.

  $13 \div 4 = 3$ rem 1 $14 \div 4 = 3$ rem 2 $15 \div 4 = 3$ rem 3

  The remainder can be 1, 2, 3, ... up to one before the number you are dividing by. Then it 'goes exactly' again. For example, $16 \div 4 = 4$

- *Sometimes, it makes sense to **round down** your result and to ignore the remainder: $13 \div 4 = 3$ (rounded down).*
  *Sometimes it makes more sense to **round up** and to allow for the remainder as if it were a whole 1: $13 \div 4 = 4$ (rounded up).*

Key facts

Rounding is a way of coping with a remainder and arriving at an answer that makes sense.

Language

In $16 \div 5$, the 16 is **divided by** the 5, and the 5 is **divided into** the 16.
Remainder: what is left over when you divide by a number that doesn't 'go exactly'.

On track

1. Do these divisions by finding how many groups you can make.
   a) 10 ÷ 5   b) 25 ÷ 5   c) 35 ÷ 5
   d) 45 ÷ 5   e) 55 ÷ 5   f) 60 ÷ 5

Aiming higher

1. Do these divisions. Each one has a remainder.
   a) 26 ÷ 4   b) 24 ÷ 5   c) 33 ÷ 6
   d) 38 ÷ 4   e) 42 ÷ 5   f) 52 ÷ 6

2. Do these divisions, and for each one round down the answer.
   a) 36 ÷ 5   b) 34 ÷ 4   c) 43 ÷ 6
   d) 58 ÷ 5   e) 52 ÷ 6   f) 22 ÷ 4

3. Do these divisions, and for each one round up the answer.
   a) 46 ÷ 4   b) 44 ÷ 5   c) 53 ÷ 6
   d) 28 ÷ 5   e) 62 ÷ 4   f) 32 ÷ 6

Using and applying

1. Imagine you have 27 eggs and that you put them into groups of 6.
How many groups are there? How many eggs are left over?

2. Imagine placing the 27 eggs in a tray
in rows of 6. How many rows are there?
Explain why.

3. Imagine putting the 27 eggs
into boxes that can hold 6 eggs
each. How many boxes do
you need? Explain why.

16 Division as the inverse of multiplication

What do you need to know?

- The multiplication number facts (and matching division facts) for the 2, 3, 4, 5, 6 and 10 times tables

What will you learn?

- How to relate a multiplication number sentence to a division sentence
- How to relate a division number sentence to a multiplication sentence

Example

This table gives lots of multiplications facts:

	2	3	4	5	6
2	4	6	8	10	12
3	6	9	12	15	18
4	8	12	16	20	24
5	10	15	20	25	30
6	12	18	24	30	36

- Look at the symmetry in the table and how matching numbers make patterns.

It does not matter in what order you multiply two numbers together. You get the same answer.
So, for each multiplication fact you can write another multiplication fact:

$$3 \times 4 = 12 \qquad 4 \times 3 = 12$$

You can also write two division facts linking the numbers 3, 4 and 12:

$$12 \div 3 = 4 \qquad 12 \div 4 = 3$$

Key facts

Multiplication is repeated addition. **Division** is repeated subtraction. Division is the inverse of multiplication. Multiplication is the inverse of division.

Language

Number sentence: a statement explaining how numbers relate to each other.

On track

1 For each of these multiplication facts write another multiplication fact.
   a) $6 \times 9 = 54$ **b)** $7 \times 6 = 42$
   c) $5 \times 4 = 20$ **d)** $4 \times 8 = 32$

2 For each of these multiplication facts, write a division fact.
   a) $6 \times 9 = 54$ **b)** $7 \times 6 = 42$
   c) $5 \times 4 = 20$ **d)** $4 \times 8 = 32$

Aiming higher

1 For each of these multiplication facts, write another multiplication fact
and two division facts.
   a) $7 \times 9 = 63$ **b)** $8 \times 6 = 48$
   c) $7 \times 4 = 28$ **d)** $5 \times 8 = 40$

2 Calculate these products. For each one, write a corresponding
multiplication fact and two corresponding division facts.
   a) 8×9 **b)** 7×8
   c) 4×6 **d)** 5×7

Using and applying

1 Write down the division calculation for this problem:
How many 5-minute cartoons could you watch in 30 minutes?
What multiplication fact can you use to find the answer to the problem?

2 *I start with the number 18. I multiply it by 5 and then divide the answer
by 5. What number do I get?* Explain how you know.

3 Make up a number problem like the one in question 2, which involves
multiplication and division. Try it on a friend.

17 Unit fractions

What do you need to know?

- The multiplication number facts (and matching division facts) for the 2, 3, 4, 5, 6 and 10 times tables
- How to record length (in metres and kilometres), weight (in grams and kilograms) and capacity (in millilitres and litres)

What will you learn?

- How to find unit fractions of numbers
- How to find unit fractions of quantities

Example

- When you share a cake between two people, each person gets a half share. The whole cake is divided into 2 pieces. This is written as $\frac{1}{2}$.

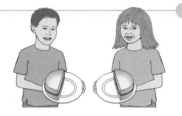

- When you share a pie between four people, the pie is cut into quarters. The whole pie is divided into 4 pieces. Each share is written as $\frac{1}{4}$.

| The 1 stands for the whole pie. | $\rightarrow$ $\frac{1}{4}$ $\leftarrow$ | The denominator shows how many shares the pie was cut into. |

- Measures can also be shared. You may convert the units for a more useful answer.

$\frac{1}{4}$ of 1 litre = $\frac{1}{4}$ of 1000 millilitres = 250 millilitres

Key facts

There are 1000 metres in a kilometre.
There are 1000 grams in a kilogram.
There are 1000 millilitres in a litre.

Language

Unit fraction: a fraction with a 1 as the numerator. This means the 1 is the top number on the fraction.

On track

1 Find these values.

a) $\frac{1}{2}$ of 18   b) $\frac{1}{2}$ of 14   c) $\frac{1}{2}$ of 12   d) $\frac{1}{3}$ of 24

e) $\frac{1}{3}$ of 18   f) $\frac{1}{3}$ of 15   g) $\frac{1}{4}$ of 12   h) $\frac{1}{4}$ of 24

2 Find these quantities.

a) $\frac{1}{2}$ of 20 cm   b) $\frac{1}{2}$ of 16 ml   c) $\frac{1}{2}$ of 10 g   d) $\frac{1}{3}$ of 21 ml

e) $\frac{1}{3}$ of 27 g   f) $\frac{1}{3}$ of 12 cm   g) $\frac{1}{4}$ of 16 g   h) $\frac{1}{4}$ of 20 cm

Aiming higher

1 Find these quantities.

a) $\frac{1}{2}$ of 4 m   b) $\frac{1}{2}$ of 8 litres   c) $\frac{1}{2}$ of 6 kg

d) $\frac{1}{4}$ of 12 kg   e) $\frac{1}{4}$ of 24 km   f) $\frac{1}{4}$ of 36 ml

2 Find these quantities.

a) $\frac{1}{2}$ of 1 m   b) $\frac{1}{2}$ of 1 litre   c) $\frac{1}{2}$ of 1 kg

d) $\frac{1}{4}$ of 1 kg   e) $\frac{1}{4}$ of 1 m   f) $\frac{1}{4}$ of 1 litre

3 One half of 36p is 18p. What is one quarter of 36p?

Using and applying

1 This line is 6 cm long. Draw a line of this length and use a ruler to divide it into quarters. _____

2 Milly has a 120 ml bottle of medicine. She takes one sixth of the medicine each day. How many days does she take the medicine for? How much medicine does she take each day? What calculation did you do to work this out?

3 Jane has a 100 g bar of chocolate. She cuts it into five equal pieces. How much does each piece weigh? What fraction of the bar is one piece?

18 2-D and 3-D shapes

What do you need to know?

- What common 2-D shapes and 3-D solids look like in different positions and orientations
- Sort, make and describe 2-D shapes, referring to their properties

What will you learn?

- How to describe and classify 3-D solids
- How to draw and make 2-D shapes

Example

- To draw a 2-D shape, first check how many sides it has. Note any special properties, such as right angles, equal sides or parallel lines. Use a ruler to draw the straight lines.

Draw shapes with right angles on squared paper.

Draw equilateral triangles and hexagons on an isometric grid.

Key facts

A **triangle** has three angles. A triangle with all three sides the same is called **equilateral**. A triangle with only two sides the same length is called **isosceles**. If all the sides are different lengths the triangle is called **scalene**.

Equilateral

Isosceles

Scalene

A **quadrilateral** has four sides. **A square** is the quadrilateral with all sides the same length and all angles the same. Curved lines can be used to create shapes: **semi circles**, **circles** and **ovals**.

Square

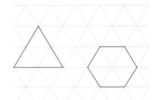

Semi circle

Circle

Oval

Language

2-D shapes are made from **lines**. The point where any two lines meet to form a corner is called a **vertex**. **3-D solids** are constructed from shapes, called **faces**, or sides. Pairs of faces join along a solid **edge**.

On track

1   Write one or more names for each shape.

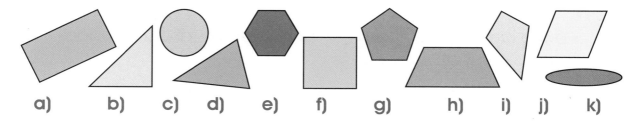

a)   b)   c)   d)   e)   f)   g)   h)   i)   j)   k)

2   Give the full name for each of these 3-D solids.

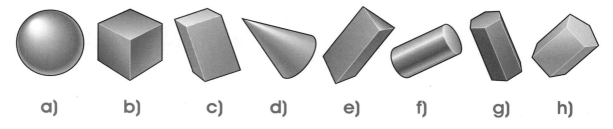

a)   b)   c)   d)   e)   f)   g)   h)

Aiming higher

1   Pair these shapes together. In each pair, one is half of the other.

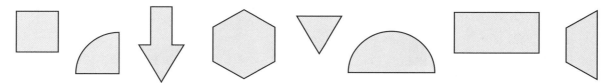

2   Spot the odd one out in these sets of shapes. Circle it and explain why it doesn't belong.

a)   b)   c)

Using and applying

1   Look around your classroom. What 2-D shapes can you see?

2   Look at containers in your kitchen cupboard. What shapes are used to hold soup, cereals or other foods? Sketch some examples.

19 Reflections in a mirror

What do you need to know?

- How to identify reflective symmetry in patterns and 2-D shapes
- How to draw lines of symmetry in shapes

What will you learn?

- How to draw and complete shapes with reflective symmetry
- How to draw the reflection of a shape on a mirror line along one side

Example

- In shapes that have reflective symmetry, one half of the shape looks like the reflection of the other half. The line of symmetry acts like a mirror line. You can use a mirror to find the line of symmetry.

- To draw the other half of a shape, plot points of the image by mirroring points in the object. Then join these points with lines to match those in the object.

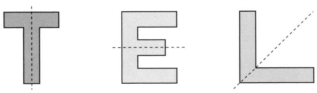

Key facts

The **image** that you see in the mirror is the same distance from the mirror line as the **object**.

Language

Line of symmetry: the line that cuts a shape in half.
Object: the shape that is reflected.
Image: the reflection of an object. It's what you see in the mirror.

On track

1 Draw one line of symmetry on each of these shapes. Use a mirror to help you.

2 Some letters of the alphabet have reflective symmetry. Use a mirror to find out which letters do then draw in the lines of reflective symmetry.

A   B   C   D   E   F   G   H   I

J   K   L   M   N   O   P   Q   R

S   T   U   V   W   X   Y   Z

3 Find the reflection of this shape.

Aiming higher

1 Copy these half-shapes and complete them to create a shape with reflective symmetry.

a   b   c

2 Copy and complete these half-letters to reveal the message.

3 Make up your own reflection message. Check it in a mirror.

Using and applying

1 Draw two straight lines to create a right angle. Move a mirror to create different quadrilateral shapes with your lines. Which of these can you make: rectangle, square, rhombus and kite?

2 Repeat question 1 but experiment by using two mirrors, held at right-angles to each other, to create new reflected shapes.

20 Position, direction and movement

What do you need to know?

- How to follow and give instructions involving position, direction and movement

What will you learn?

- How to use the four compass directions to describe movement
- How to read and record position, direction and movement

Example

- On this map, the school is north of the post office. The railway is west of the library. The library is south of the post office.

Key facts

The sun rises in the east and sets in the west. A compass points north.
Position describes where you can find something. It can show the position in relation to a grid or how far away something is from a particular point.
Direction is a measure based on angles. It can be given in terms of compass settings, such as north of you, south-west of you, or in relation to where you are now, such as ahead and behind.

Language

Clockwise: the same direction as the hands on a clock move. **Anticlockwise:** the opposite direction.

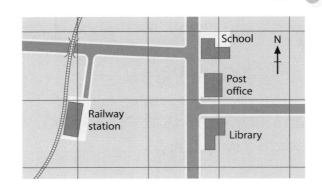

Clockwise   Anticlockwise

Movement: a change in position or direction. It can be given in terms of a shift, such as two steps forward, or a turn, such as a turn through a right-angle clockwise.

 **On track**

1 Imagine you are facing north before each of these turns.
Which direction would you be facing after the turn?

a) A $\frac{1}{4}$ turn clockwise

b) A $\frac{1}{2}$ turn anticlockwise

c) A $\frac{3}{4}$ turn anticlockwise

d) A $\frac{3}{4}$ turn clockwise

2 Imagine you are facing west before each of these turns.
Which direction would you be facing after the turn?

a) A $\frac{1}{4}$ turn clockwise

b) A $\frac{1}{2}$ turn anticlockwise

c) A $\frac{3}{4}$ turn anticlockwise

d) A $\frac{3}{4}$ turn clockwise

 Aiming higher

1 **a)** Put an X in the square that is 2 squares
north of 3B.

b) Start at 3B again. Go 2 squares east,
and then 4 squares north.
Mark your spot with the letter Y.

c) Start at 3B again. Go 1 square south, then
2 squares west. Mark your spot with the letter Z.

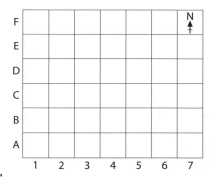

2 Make up your own route, from 1A to 6F. Test it on a friend.

 Using and applying

1 Choose two rooms in your house, such as your bedroom and the
kitchen. Give directions on how to get from one to the other. Test out your
instructions. Did they work? Or would you have bumped into a wall?

2 Choose two locations on a local map, such as your school and your
home. Give directions on how to get from one place to the other. Compare
your instructions with those given on the Internet.

21 Right angles in 2-D shapes

What do you need to know?

- That a right angle stands for a quarter turn
- How to recognise 2-D shapes in different positions and orientations

What will you learn?

- How to use a set square to identify right angles in 2-D shapes
- How to compare angles with a right angle
- How to recognise that a straight line is equivalent to two right angles

Example

- A set square has one right angle. A right-angled triangle also has one right angle.

- A square has four right angles.

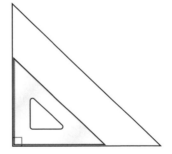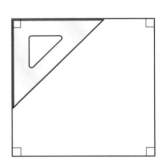

Key facts

Angle: a measure of turn, measured in degrees (°), from 0° (no turn) to 360° (a full turn).

Acute angle: smaller than a right angle.
Obtuse angle: bigger than a right angle but smaller than two right angles.
A **straight line** is equivalent to two right angles.

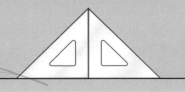

Language

Diagonal: a line drawn across a shape from one vertex (or 'corner') to another vertex.

On track

1 Use a set square to check these angles.

Which are smaller than a right angle?

Which are right angles?

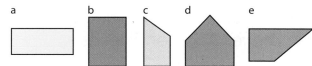

Which are bigger than a right angle? Label the angles as acute, right or obtuse.

2 Use a set square on the angles in these 2-D shapes.

Mark the ones that are right angles.

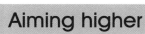

Aiming higher

1 Which of these angles are smaller than a right angle?

Use your set square to check.

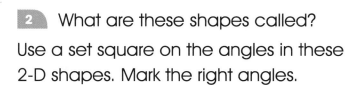

2 What are these shapes called?

Use a set square on the angles in these 2-D shapes. Mark the right angles.

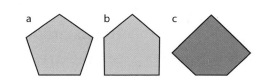

3 What are these shapes called?

Mark the right angles.

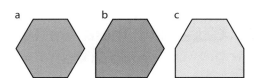

Using and applying

1 Look at the diagonals of this square.

Do they cross at right angles?

2 Look at the diagonals of this rhombus.

Do they cross at right angles?

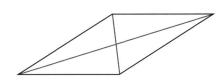

22 Units of length, weight and capacity

What do you need to know?

- Units of measure: metres, grams and litres

What will you learn?

- The relationship between kilometres and metres, and between metres and centimetres
- The relationship between kilograms and grams
- The relationship between litres and millilitres

Example

- Length: **There are 100 centimetres in 1 metre.**
  So 200 cm = 2 m, 300 cm = 3 m, 400 cm = 4 m, ...
  There are 1000 metres in a kilometre.
  So 2000 m = 2 km, 3000 m = 3 km, 4000 m = 4 km, ...

- Weight: **There are 1000 grams in a kilogram.**
  So 2000 g = 2 kg, 3000 g = 3 kg, 4000 g = 4 kg, ...

- Capacity: **There are 1000 millilitres in a litre.**
  So 2000 ml = 2 litres, 3000 ml = 3 litres, 4000 ml = 4 litres, ...

Key facts

Metres, **centimetres** and **kilometres** are used to measure length.
Grams and **kilograms** are used to measure weight.
Litres and **millilitres** are used to measure capacity.
Capacity is a measure of how much something can hold, e.g. how much liquid there is in a bottle.

Language

cm is short for centimetre. **Centi-** means a one-hundredth part.
ml is short for millilitre. **Milli-** means a one-thousandth part.
kg is short for kilogram. **Kilo-** means one thousand times.

On track

1 Complete these statements:
   a) 500 cm = ... m
   b) 6000 m = ... km
   c) 6000 g = ... kg
   d) 5000 ml = ... litres

2 Complete these statements:
   a) 4 m = ... cm
   b) 2 km = ... m
   c) 3 kg = ... g
   d) 7 litres = ... ml

3 Complete these statements. The first one has been done for you.
   a) $\frac{1}{2}$ kg = 500 g
   b) $\frac{1}{2}$ litre = ... ml
   c) $\frac{1}{2}$ m = ... cm

   d) $\frac{1}{2}$ km = ... m
   e) $\frac{1}{4}$ kg = ... g
   f) $\frac{1}{4}$ litre = ... ml

Aiming higher

1 Which unit would you use to weigh each of these items?
   a) a horse   mg / g / kg
   b) a chicken's egg   mg / g / kg
   c) a leaf   mg / g / kg

2 Which unit would you use to measure these items: mm, cm, m or km?
   a) thickness of a pen nib
   b) diameter of a football
   c) distance to school
   d) width of a room

Using and applying

1 Kate has a 1 metre length of ribbon. She cuts it in half, then in half again to make four shorter ribbons. How long is each ribbon, in centimetres?

2 Kate puts 1 kg of flour into a bowl. The recipe asks for half as much fat. How much fat should Katie add, in grams?

3 Kate empties two cartons of juice into a 1 litre jug. Each carton contains 200 ml of juice. How many more cartons of juice would be needed to fill the jug?

23 Reading scales

What do you need to know?

- How to read the numbered divisions on a scale and how to interpret the divisions between them
- How to use a ruler to draw a line to the nearest centimetre

What will you learn?

- How to interpret scales that are partially numbered
- How to read scales to the nearest division and half-division
- How to draw a line to a suitable degree of accuracy

Example

- Often scales are only partially numbered.

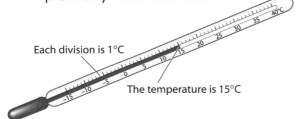

Each division is 1°C

The temperature is 15°C

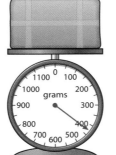

Each division is 50 g. The arrow points halfway between 400 g and 450 g so the parcel weighs 425 g.

- When you measure between numbered marks, count the number of divisions to decide how to interpret what each mark means.

Key facts

There are 1000 millimetres in a metre and 100 centimetres in a metre, so there are **10 millimetres in a centimetre**.

Language

Degree of accuracy: this is the appropriate unit of measurement for different kinds of things. For example, age is measured in years, a large piece of paper is measured in centimetres and a small insect is measured in millimetres, while athletic races are measured in metres and travelling distances in kilometres.

 **On track**

1 Identify these measurements in millimetres.

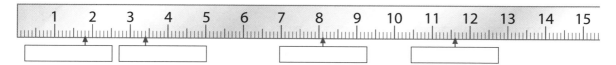

2 Measure these lines using a ruler. Give your answers to the nearest half centimetre.

 Aiming higher

1 Measure the height and width of this page. What units will you use?

2 Weigh this book. What units will you use? Estimate how much 10 of these books would weigh. Weigh 10 books and check your estimate.

3 Compare your answers to questions 1 and 2 with others in your class. Did you use the same units? Do you have the same answers?

Using and applying

1 Draw this shape using a ruler to make sure you have the correct measurement. Use a set square to make sure the angles are all right angles.

2 Use a street map or Ordnance Survey map to measure the distance from your house to your school approximately, or 'as the crow flies'. Give your answer in kilometres, to the nearest kilometre. Compare this with how far it is by road.

24 Intervals of time

What do you need to know?

- Units of time: seconds, minutes, days and hours
- How to read the time to the nearest quarter hour
- How to identify time intervals, including ones that cross the hour

What will you learn?

- How to read the time on a 12-hour digital clock
- How to read the time on an analogue clock to the nearest 5 minutes
- How to calculate time intervals
- How to find start times and end times for a time interval

Example

- There are two ways of displaying time: digitally and on an analogue clock.

On the digital clock, the time is given in numbers.

The hours (from 1 to 12) The minutes (from 0 to 59)

On the analogue clock, two hands show the time.

The short hand points to the hour.

The long hand points to the minutes.

Key facts

When an analogue clock says 9:30, you cannot tell if it means in the morning or evening. The same goes for a 12-hour digital clock. When you write the time, to show you mean in the morning, write **a.m.** (ante meridian). To show that that the time is in the afternoon, write **p.m.** (post meridian).

9:30 a.m.   9:30 p.m.

Language

Analogue clock: shows how time is always passing. The hands move smoothly. **Digital clock:** shows the time to the nearest minute. It 'clicks' from one time to the next.

On track

1   Complete this statement: There are .... seconds in 1 minute, ... minutes in 1 hour, and .... hours in 1 day.

2   Copy and label this analogue clock.

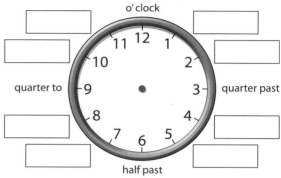

o'clock

quarter to

quarter past

half past

3   Write these times in this form: 12:15.

a)   b)   c)   d)

 Aiming higher

1   Copy and draw the hands on these clocks for these times.

a) 10:20   b) 8:15

c) 3:45   d) 12:30

2   What time does a 12-hour digital clock show at noon?
    What time does a 12-hour digital clock show at midnight?

 Using and applying

1   These clocks show the time Ali boards a bus and the time he arrives at his destination.
How long is his journey?

2   These clocks show the time a lesson starts and the time it finishes.
How long is the lesson?

25 Tally charts and frequency tables

What do you need to know?

- How to collect and record data in lists and tables

What will you learn?

- How to collect, organise and present data in tally charts
- How to create a frequency table

Example

Kate throws a die 50 times and records this in a table.

1	4	5	2	6	5	6	1	2	2	2	6	2	2	5	4	2	1	3	6	1	3	5	5	2
3	6	1	4	3	1	3	6	5	5	5	3	4	1	1	4	3	5	4	1	5	4	6	3	4

It is hard to make sense of this data. She transfers it to a tally chart.

The frequency data can be entered into a spreadsheet and a bar chart created.

Score	Tally	Total
1	1111 1111	9
2	1111 111	8
3	1111 111	8
4	1111 111	8
5	1111 1111	10
6	1111 11	7

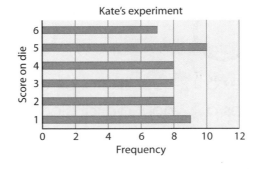

Kate's experiment

	A	B
1	Score	Frequency
2	1	9
3	2	8
4	3	8
5	4	8
6	5	10
7	6	7

Key facts

On a **fair die**, each score is equally likely to happen.

Language

Frequency: the number of times an outcome, such as throwing a six, happens.

 **On track**

1 Use this data to create a tally chart.

1	5	6	6	2	2	2	5	2	3	1	5	2	1	4
3	1	3	3	5	5	4	1	3	4	5	6	4	2	5

Calculate the frequency for each value 1 to 6.

Check that the total frequency is 30.

2 Look at this data. How many times was the die thrown?

3 Enter the data in question 2 on to a spreadsheet, then create a bar chart.

4 Enter your data from question 1 on a spreadsheet, then create a bar chart.

Score	Frequency
1	5
2	7
3	6
4	7
5	7
6	8

 Aiming higher

1 Repeat Kate's experiment. Record your scores in a rectangular table so you know when you have thrown the die enough times.

2 Transfer your data on to a tally chart. Calculate the frequency for each score, then check that they total 50.

3 Enter your data into a spreadsheet then create a bar chart.

 Using and applying

1 How could you use the data collected from your experiment to decide whether the die was fair?

2 How many times should you throw a die before you have enough data to make a judgement on whether it is fair or not?

26 Using diagrams to sort data

What do you need to know?

- How to use lists, tables and diagrams to sort objects
- The language of sorting, including 'not'

What will you learn?

- How to use a Venn or a Carroll diagram to sort data and objects

Example

- A **Carroll diagram** can be used to sort data according to two criteria.

Numbers 1 to 15	Multiple of 2	Not multiple of 2
Multiple of 5	10	5, 15
Not multiple of 5	2, 4, 6, 8, 12, 14	1, 3, 7, 9, 11, 13

- A **Venn diagram** can be used to categorise data too.

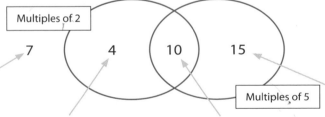

7 is outside both circles. It is not a multiple of 2 and it is not a multiple of 5.

4 is a multiple of 2, but not a multiple of 5.

10 is in the overlapping region. It is a multiple of 2 and a multiple of 5.

15 is a multiple of 5 but not a multiple of 2.

Key facts

A **Venn diagram** with two circles overlapping creates four regions. An object either belongs inside a circle, or not.
A **Carroll diagram** with two rows and two columns creates four areas.

Language

Criterion: a way of sorting, e.g. by multiples of 5 or by multiples of 2.

 **On track**

1 Copy the Venn diagram in the Example and place the numbers 1 to 15 in the correct places.

2 Copy the Carroll diagram in the Example and extend it to show the numbers from 1 to 20 which are multiples of 2 and multiples of 5.

Aiming higher

1 Create a Carroll diagram to identify the numbers from 1 to 20 that are multiples of 3 and those which are multiples of 5.

2 Create a Venn diagram to identify the numbers from 1 to 20 that are odd, and those which are multiples of 5.

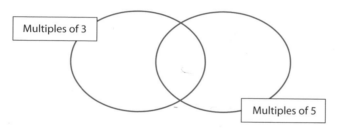

3 Look at this Carroll diagram.

Fruits	Pips	No pips
Can eat the peel	apple, grape	cherry, plum
Can't eat the peel	orange, lemon	banana, pineapple

Present the same information using a Venn diagram.

 Using and applying

1 Compare the shoes you are wearing with others in your class. Choose two criteria, such as the colour and whether the shoe has laces. Use a Carroll diagram to show your findings.

2 Ask others in your class to find out who has brothers and who has sisters. How many have both brothers and sisters? Use a Venn diagram to show your results.

27 Problem solving

What do you need to know?

- How to solve problems about numbers, measures or money

What will you learn?

- How to solve one-step and two-step problems, choosing and carrying out appropriate calculations

Example

- Problems are written using words which tell you the operation to use.

| **'Altogether'** means you will have to use addition. | Jane has two 50p coins and a 20p coin. How much is this **altogether**? | There are two 50p coins, so you need to multiply. |

The problem will also tell you which numbers to use: 2 and 50 and 20.

Some calculations you can do in your head. For others, you need to record your calculations.

You might write $(2 \times 50) + 20 = 100 + 20 = 120$

You may have to make your answer fit the problem. It is about money, so you must include the units. You might also write £1.20 instead of 120p.

Key facts

All the information you need is written in the question. Read it carefully. To be double sure, read it twice, or even three times.

Language

Problem solving: means working out what to do to find the answer.

On track

1 Look at this problem. *Ella buys a 10p lolly. She pays with a 50p piece. How much change does she get?*
Which calculation will you use to solve this problem?

$$50 + 10 \qquad 50 - 10 \qquad 50 \times 10 \qquad 50 \div 10$$

How did you choose the correct calculation? What unit is the answer in?

2 Look at this problem. *Three buns cost 24p. What do two buns cost?*
What numbers will you use to solve this problem? Write down the calculations you need to do.

Aiming higher

1 Look at this problem. *I buy two comics that cost 45p each. How much change will I get from £1?*
I start to solve this problem by writing $2 \times 45 = 90$.

What does 90 tell me? What do I have to do next to solve the problem?

2 Look at this problem.

Ling got into the swimming pool at 10:30. She got out at 11:20. How long was she in the pool?
What numbers will you use to solve this problem?

What fact, linking hours and minutes, do you need to know?

Using and applying

1 What is the answer to 20×5? Make up a problem that uses the calculation 20×5. How do you recognise that this problem involves multiplication?

2 Make up your own word problem that uses the calculation $32 \div 4$. How do you recognise that your problem involves division?

28 Using images and diagrams to solve problems

What do you need to know?

- How to identify and record information needed to solve a problem

What will you learn?

- How to use numbers, images and diagrams to find a solution
- How to present a solution in a appropriate way

Example

- Some problems can be solved in your head. For others, you might need to write something down, such as a **number sentence**.

$$3 + 4 + 5 + 6 + 7 = 3 + 7 + 4 + 6 + 5$$
$$= 10 + 10 + 5$$
$$= 25$$

A number line might help you to explain a calculation, such as $25 + 9$.

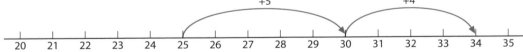

How many legs do 3 zebras have? A picture may help you to count.

Jack is 120 cm tall. His sister is 97 cm tall. How much shorter is she?

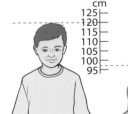

The difference in heights shows how many cm taller Jack is than his sister.

Key facts

Numbers can be added in any order. The answer will be the same.

Language

Number sentence: a mixture of numbers and operations, such as addition and multiplication.

Words such as **shorter** and **longer** or **taller** tell you to work out the **difference** between two lengths.

On track

1 Look at this problem. *I think of a number and add 7 to it. My total is 13. What is my number?*
How would you start to do this problem? Use a number line to solve it.

2 Look at this problem. *How many wheels are there on 4 bicycles?*
Draw a picture to show how you worked out your answer.

3 Look at this problem. *Two snakes are 49 cm and 62 cm long. What is the difference in their lengths?* Draw a diagram to help you to solve the problem. What part of your diagram shows the answer?

4 *Sam had 50 pies. He sold some and then had 20 left.*
Which of these number sentences shows this?

A ? − 20 = 50   **B** 20 + ? = 50   **C** ? − 50 = 20   **D** 50 − ? = 20

Aiming higher

1 Look at this problem. *Jo has five £1 coins and two 10p coins in her purse. How much does she have altogether?* Draw a picture to show how much money Jo has in her purse.

2 Look at this problem. *12 grapes are shared equally on to 4 plates. How many grapes are there on each plate?*
What calculation would you do to find the answer? Draw a picture to represent the problem.

3 Look at this problem. *How many bunches of 4 grapes can you get from 12 grapes?* What calculation would you? Draw a picture of this problem.

Using and applying

1 Think of a problem that this picture might show.

2 Think of a problem that this picture might show.

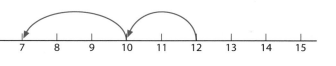

29 Following lines of enquiry

What do you need to know?

- How to answer questions by choosing and using suitable equipment

What will you learn?

- How to identify what information is important
- How to make and use lists, tables and graphs to organise and interpret information

Example

- To follow a line of enquiry, you ask questions and find answers to these questions. The questions you ask will depend on what you want to know. So first ask yourself what are you trying to find out?

Then you need to decide what data you need to collect and how to collect it.

- To find out information about your schoolmates, you might ask everyone to complete a **questionnaire**. For other problems, you might do some calculations.

There are lots of ways of recording the results. You can write a list. Or, if you have lots of data, you can create a table, or draw a pie chart or some other diagram or graph.

Key facts

Data is what you collect. When you analyse your data and present your findings, you create **information**.

Language

Questionnaire: a list of questions used to find out some information.

On track

1 Think about this problem. *You want to know how much colder it is at night compared with the temperature at midday.*
How can you collect this data? How can you record it? How can you present your findings?

2 Think about this problem. *You want to know what pets your classmates have.*
How can you collect this information? How can you record it? What type of diagram might best show your results? How would you be able to see from your results which pet was the most popular type of pet?

Aiming higher

1 Think about this problem. *In a hospital, the nurses need to keep a check on the temperature of a patient who has a fever.*
How could the nurses collect this information? How could they record it? What might the record show?

2 Think about this problem. *A newsagent provides a delivery service for daily and weekend papers to customers who live within one mile of the shop.*
How could the newsagent collect information about which papers are needed? How could the newsagent make sure the right papers are delivered to the correct homes?

Using and applying

1 Ask everyone in your class a question about their home life, e.g. how many brothers do they have? What's their favourite TV programme? How far do they live from school? Record your findings and create a chart to show the results. Did anything surprise you about the results?

2 Think about this problem. *You think that the only possible remainders when dividing by three are 0, 1 and 2.*
What could you do to check this? How would you record your results?

30 Patterns

On track

1   Look at this problem. *Sort the numbers 1 to 20 into two groups:*
'multiples of 5' and 'not multiples of 5'.
What is special about the numbers in the 'multiples of 5' group?
Give two examples of multiples of 5 which are bigger than 100.

2   Look at this problem: *8 – 2 = 6. What are 80 – 20, and 800 – 600?*
How do you know the answers? What do you understand by the zeroes in
each number?

3   Look at this problem. *In this set of 2-D*
shapes, identify the shapes that have one
right angle, two right angles, more than
two right angles.
How can you recognise a right angle?

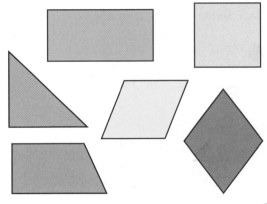

Aiming higher

1   Look at this problem. *What is the next calculation in this pattern?*
857 – 800 = 57   857 – 700 = 157   857 – 600 = 257

Explain how you know the next calculation.

2   Look at this problem. *All the shapes*
on this table except one are prisms. Which
shape does not belong? How can you
recognise the odd one out?
What are you looking for?

Using and applying

1   What are the missing numbers in this pattern? 93, 88, ..., ..., 73, 68.
How did you find them? Write another missing number problem and try it
on a friend.

2   What addition could you use to work out 15 – 7? Why can you use
addition to work out a subtraction?